Tools for Learning

Tools for Learning is an on-line review area that provides a variety of activities designed to help students study for their class. Students will find chapter outlines, review questions (written by the author), flash cards, figure labeling exercises, and links to a variety of tutorials and other useful learning tools to help students master the basic science material.

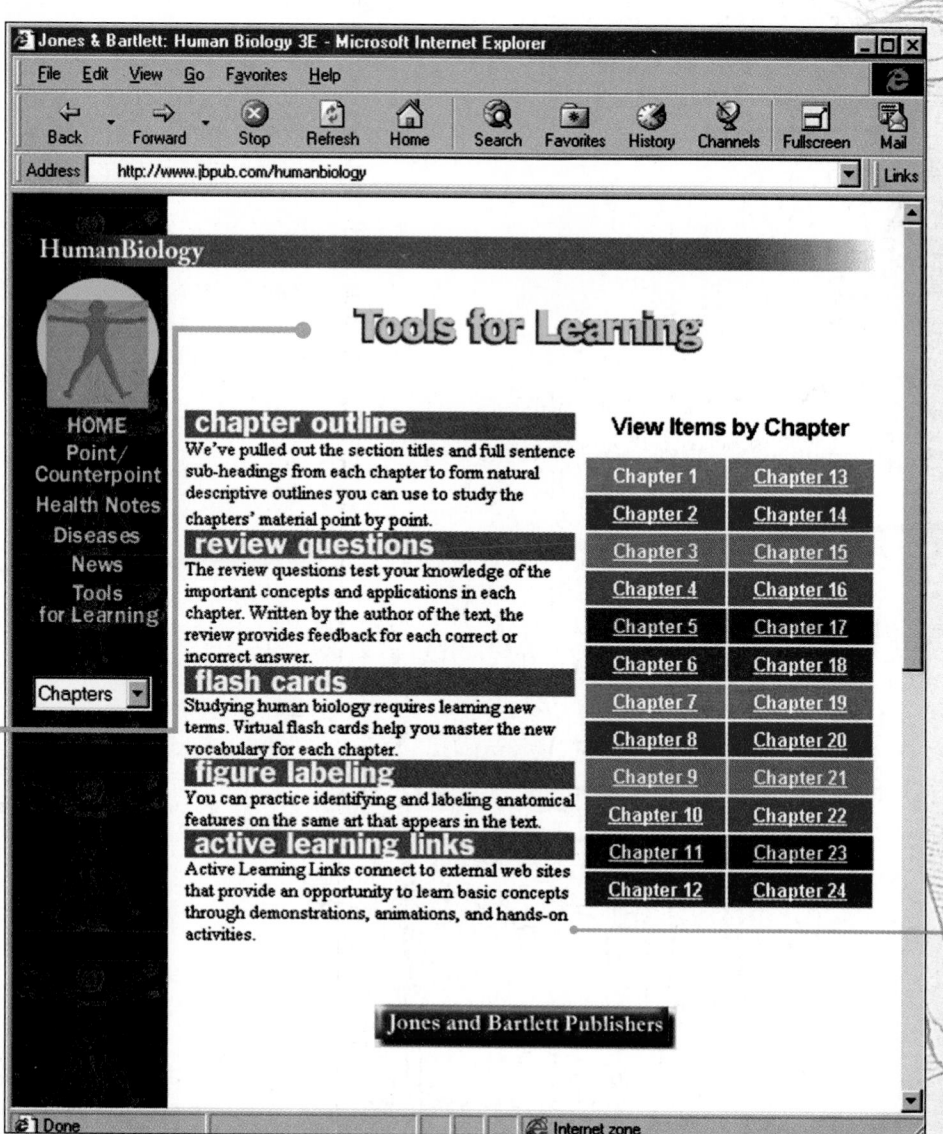

Jones & Bartlett: Human Biology 3E - Microsoft Internet Explorer

File   Edit   View   Go   Favorites   Help

Back   Forward   Stop   Refresh   Home   Search   Favorites   History   Channels   Fullscreen   Mail

Address   http://www.jbpub.com/humanbiology   Links

HumanBiology

HOME
Point/
Counterpoint
Health Notes
Diseases
News
Tools
for Learning

Chapters ▼

Tools for Learning

chapter outline
We've pulled out the section titles and full sentence sub-headings from each chapter to form natural descriptive outlines you can use to study the chapters' material point by point.

review questions
The review questions test your knowledge of the important concepts and applications in each chapter. Written by the author of the text, the review provides feedback for each correct or incorrect answer.

flash cards
Studying human biology requires learning new terms. Virtual flash cards help you master the new vocabulary for each chapter.

figure labeling
You can practice identifying and labeling anatomical features on the same art that appears in the text.

active learning links
Active Learning Links connect to external web sites that provide an opportunity to learn basic concepts through demonstrations, animations, and hands-on activities.

View Items by Chapter

Chapter 1	Chapter 13
Chapter 2	Chapter 14
Chapter 3	Chapter 15
Chapter 4	Chapter 16
Chapter 5	Chapter 17
Chapter 6	Chapter 18
Chapter 7	Chapter 19
Chapter 8	Chapter 20
Chapter 9	Chapter 21
Chapter 10	Chapter 22
Chapter 11	Chapter 23
Chapter 12	Chapter 24

Jones and Bartlett Publishers

Done   Internet zone

Each Active Learning Link is carefully introduced so that students know what they will be doing and seeing at the site and how it relates to the chapter they've been reading.

To find out more about HumanBiology, please e-mail info@jbpub.com, or call your Jones and Bartlett sales representative at 800-832-0034.

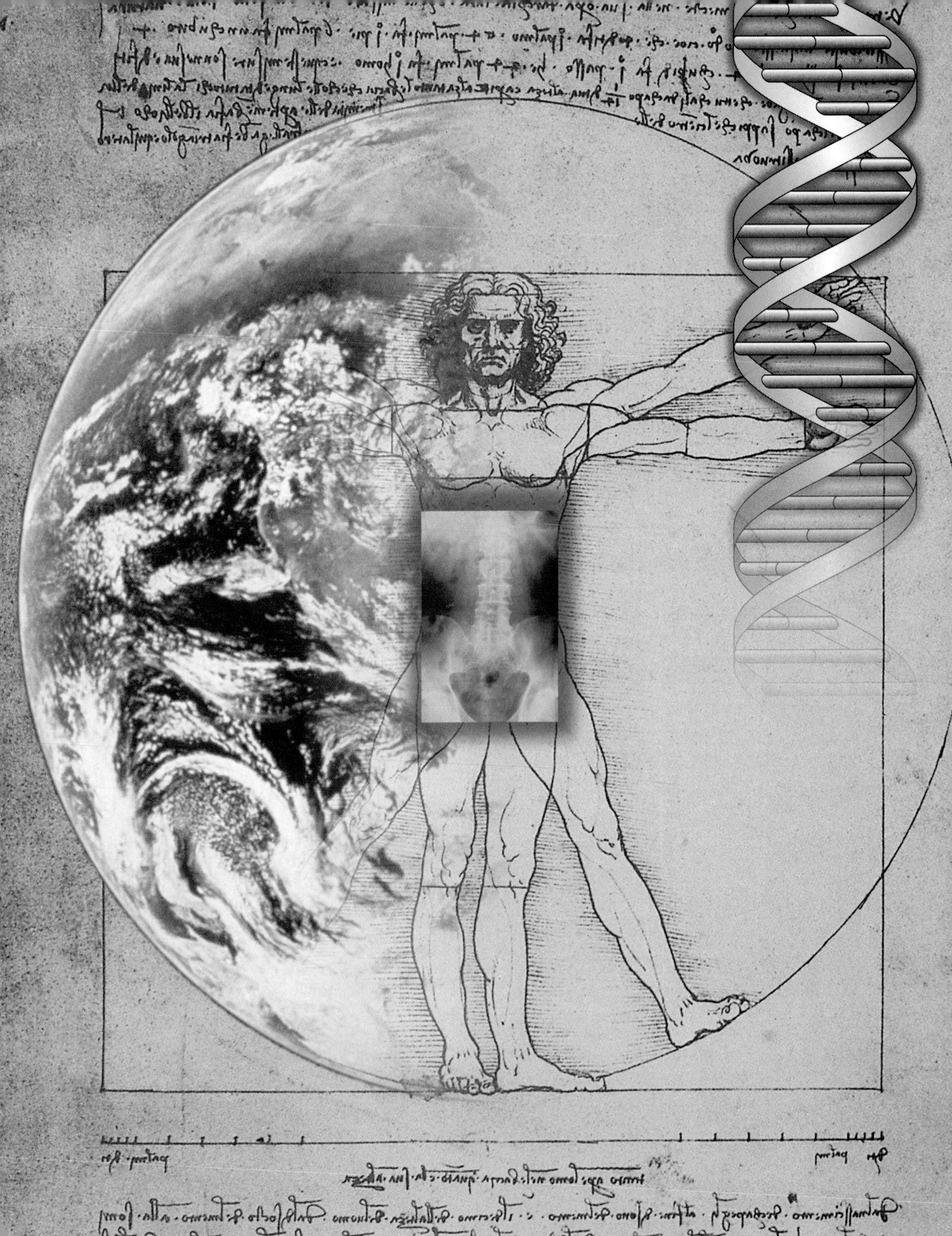